Ogi Orange
THE BIG RACE

© 2023 Lee Chau

ISBN 979-8-9875895-1-9

Published by Pybabi Publishing

This book is dedicated to my mother and father to whom my love and honor belongs.

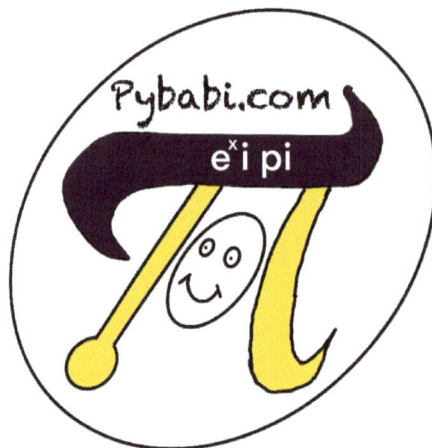

Message for instructor, teacher, or parent:

This book is best used by allowing one of the children who can read to lead the group after you have done it first as a model for them to imitate. Then they should read the story or section, ask the questions in the MATCH and ANSWER sections to their fellow students or if in a family to their brothers and sisters. As they lead the group you can monitor their progress. As the monitor you should answer any questions that arise naturally from the discussion. After one child has taken the lead, the role of teacher should be rotated to another child and then repeated until each child has successfully taught the material before advancing to another book. Since they are learning to teach the material they will naturally pay sufficient attention to learn the material and the lessons will stick with them. As the monitor, it may be best to intervene only as necessary allowing the children to solve and figure things out on their own until they need assistance.

If the child cannot read the parent can read and explain the lesson being taught. As soon as the child can understand even the simplest principle or idea have them to explain it back to you as if you are the student and they are the teacher.

When children feel they have an important role in education they will naturally learn what is necessary to play that role.

Of course this is not the only way to use the material, as you help the student you will soon discover which methods are best depending on the age and ability of the child. Please note that the illustrations in the book are specifically designed not to be perfect. They are only the beginning of a bigger idea or concept that will later be explained in greater detail. By keeping this in mind you will not have to push the student into trying to understand everything at the same time.

Thank you for choosing to use this material.

pybabi.com

OGI ORANGE

THE BIG RACE

Part I

Written and Illustrated by Lee Chau

OGI ORANGE

THE BIG RACE

Part II

Written and Illustrated by Lee Chau

OGI ORANGE

THE BIG RACE

Part III

Written and Illustrated by Lee Chau

OGI ORANGE

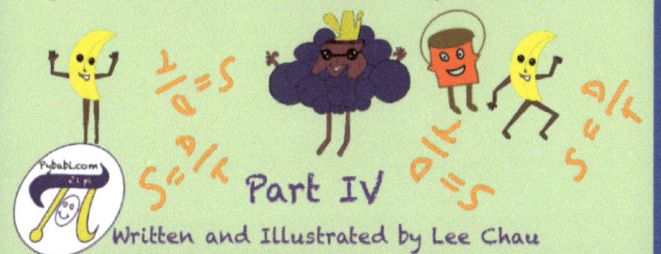

THE BIG RACE

Part IV

Written and Illustrated by Lee Chau

This is my book

Date: _____

Name:

_____

I love you Space

Please sign,
date and leave
words of
encouragement

I love you Space

Please sign, date and leave words of encouragement

I love you Space

Please sign, date and leave words of encouragement

What is average speed?

Let's find our average speed to see 'who is the fastest fruit in the valley?'

Ok, but I must get The Secret to Speed! After we find out the average speed I have to defeat Bender Pretender!

1

What is average speed?

Is it in our body?

Or under our peels?

2

What is average speed?

Is it in our body?

Or under our peels?

2

No, no, no my friends, it is a comparison. That is when we look at two things to see how they act when they are together.

We call that 'how they relate to one another' and for short we call it a relationship.

look at comparison

how they relate to one another

3 Relationship

What is a relationship?

What is a relationship?

Relationship?

Relationship?

Relationship?

4

What is a relationship?

A relationship is an agreement between two things.

Air   Balloon

5

The air and balloon are in a relationship.

Fill me up to 100.

Air

Ok, I have 5 buckets I can use.

We use our imagination that we can fill buckets with air.

6

So, is it an agreement between the two things?

Air and the balloon?

Thing

Property

Relationship

7

No, no, no, it is the agreement between their changing properties.

The balloon changing property is connected to air's changing property. As the agreement gets fulfilled the balloon 'gets larger'.

So it is the change property of each thing and not the things themselves .

8

Air

Balloon

air changing

More

Less

Size of Balloon Changing

Bigger
or
Smaller

This link is what we call a relationship

Change in amount of air

Change in size of Balloon

9

Link

Anything that has a changing property can be in a relationship or linked with something else that has a changing property.

What is a property?

Is a property some part of us?

Property?

Property?

Property?

Properties are things that make you what you are. Everything that is something has properties.

Ogi Orange is a thing.

Properties   Type
Color = Orange
Shape = Circle
Body = juicy

Balii Banana is a thing.

Properties   Type
Color = Yellow
Shape = Half circle
Body = mushy

Some things have changing properties. That means properties that can increase or decrease naturally or depending upon something outside of it self. That is a connection or a link to another changing property of something else.

Air is a thing

A balloon is a thing

Properties	Types
Color =	Invisible
Shape =	none
Amount =	more or less

Properties	Types
Color =	any color
Shape =	round
Size. =	larger or smaller

Did you notice that both "Air and Balloon" have changing properties?

What are they?

12

Did you notice that
the amount property
could be 'more or
Less'?

Did you also notice that
the size property could be
'more or less'?

Sometimes the "amount" property
only increases. That means the
thing can only become more but
never becomes Less.
"*more or less*" can also mean
"*increase or decrease*".

13

Changing properties like 'amount and size' are measurable so we can use numbers to see how many times air was added to the balloon.

Air

Units

5

Balloon

Size

100

m

V_{av}

S

$f'(x)$

But what this do not show is how MUCH air was added each of the 5 times or CHANGES until the limit of 100 was reached. Only when we DIVIDE the link can we see the AGREEMENT that is 'each time of change' of both the amount and size. So we can call this relationship an AGRREEMENT or 'each time of change', 'unit change', 'rate of change', 'derivative', and finally 'slope'. If the agreement refers to moving things then it can be called 'average speed' which is something different than average velocity but both are ratios. We can use the symbols 'Vav' = average velocity and 'S' = average speed. They are both ratios or agreements.

$\dfrac{dy}{dx}$

Air		Balloon
Units	Amount	Size change
First delivery out of 5	20	20
Second delivery out of 5	20	40
Third delivery out of 5	20	60
Fourth delivery out of 5	20	80
Fifth delivery out of 5	20	100

14

What is divide?

Is it easy or hard?

Changing Properties?

Measurable?

Divide?

15

Many times "divide" is a fancy way of saying share equally.

Numbers are invisible objects in our head. They are stored there automatically just like words

Let's say we want to share 4 objects into two containers. How many objects may go into each container?

2

2

2 objects in each container.

16

Now that we know how to divide. Lets go back and look at (compare) our "amount and size" relationship.

Balloon Total Size 100

100 Objects share by 5

Unit total 5

Balloon grows in size 20, for every 1 of the 5 changes. Each of the 5 changes can be considered as 1 unit. So the balloon received 20 for 1 unit change in air.

$$\frac{100}{5}$$

100 objects

The fascinating part about this is that once you see the link you can detach it from its original things/objects and apply it to any two objects or things or even to a single object or thing as long as the thing has both properties.

Air   Balloon
amount O————O Size

Relationship detached

To apply to a person or someone's bank account.

amount O————O Size

Single Object
Person

Food   Stomach
amount O————O Size

Single Object
Person

Money   Bank Account
amount O————O Size

18

Be patient my friends! First we must review everything you have learned..

Match the answers to the questions

MATCH 1

Non changing and changing.

What is a property?

A property is that which makes a thing. It is what makes you are. what you are.

What is a thing?

What type of properties are there?

Anything that can be talked about. Objects in our mind or things that we can sense or measure.

MATCH 2

Is the measurement of the changing properties the total measurement or only a piece of it?

What is the link called that attach the changing properties together?

1. A link
2. changing property on each side.
3. The instructions to DIVIDE.

A unit, or an increment.

What is the difference between a unit and the total measurement of a changing. property?

A unit tells us WHEN a change occurred and the total tells the what was agreed on.

A relationship.

Total

Amount and Size

What are the parts of a relationship?

What are some types of changing properties?

What do we call a change is made? 'each time

MATCH 3

A property is a common group like color. A type is a specific type of the common group. It answers the question: what is the exact color? Type= red.

How one changing property CHANGES with each unit change of another changing property.

What does a relationship show?

To share equally, to distribute equally.

Is 'time' a non changing property or is it a changing property?

Since distance can be measured it is a changing property.

If 'time' is a changing property, does it only increase that is move forward or does it also move backwards?

Time only increases that means it only moves forward.

Since time can be measured it is a changing property.

Is 'distance' a non changing property or is it a changing property?

What is the difference between a property and a type?

What is one definition of 'to DIVIDE'?

ANSWER 1

Is the measurement of the changing properties the total measurement or only a piece of it?

What do we call 'each time a change is made?

What is the link called that attach the changing properties together ?

What is a thing?

What is the difference between a unit and the total measurement of a changing. property?

What type of properties are there?

What are the parts of a relationship?

What is a property?

What are some types of changing properties?

Try to answer these on your own!

ANSWER 2

Is 'distance' a non changing property or is it a changing property?

What is the difference between a property and a type?

If 'time' is a changing property, does it only increase that is move forward or does it also move backwards?

Is 'time' a non changing property or is it a changing property?

What does a relationship show?

What is one definition of 'to DIVIDE'?

ANSWER 3

Try to answer these on your own.

What is one definition of the word divide?

What does a relationship show?

Is 'distance' a non changing property or a changing property?

Is 'time' a non changing property or a changing property?

If 'time' is a changing property does it only increase that is move forward or does it also decrease that is move backwards?

What is the difference between a property and a type?

Teachers Space

Please sign, date and leave words of encouragement

Friends Space

Please sign,
date and leave
words of
encouragement

Mess up Space

Do whatever you want here.

OGI ORANGE

THE BIG RACE

Part I

Written and Illustrated by Lee Chau

OGI ORANGE

THE BIG RACE

Part II

Written and Illustrated by Lee Chau

OGI ORANGE

THE BIG RACE

Part III

Written and Illustrated by Lee Chau

OGI ORANGE

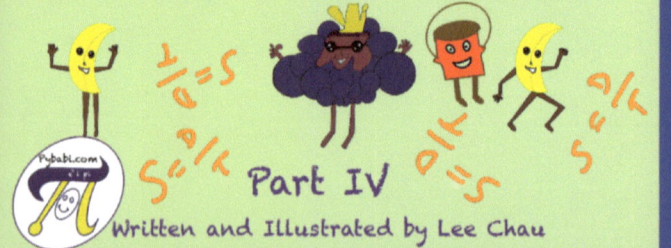

THE BIG RACE

Part IV

Written and Illustrated by Lee Chau

Let's go to part III

Can you share your story?

Tiktok @ogi.orange

Twitter @ogi_orange

Instagram @ogi.orange

Email: ogiorange@pybabi.com

www.ingramcontent.com/pod-product-compliance
Lightning Source LLC
Chambersburg PA
CBHW041549040426
42447CB00002B/102